P9-CQZ-502

Grade 1

KUMON MATH WORKBOOKS

Word Problems

Table of Contents

KUMON

Addition

1

Level ★

Score

/100

Date / /

Name

1 How many coins are there in all?　　　　　10 points

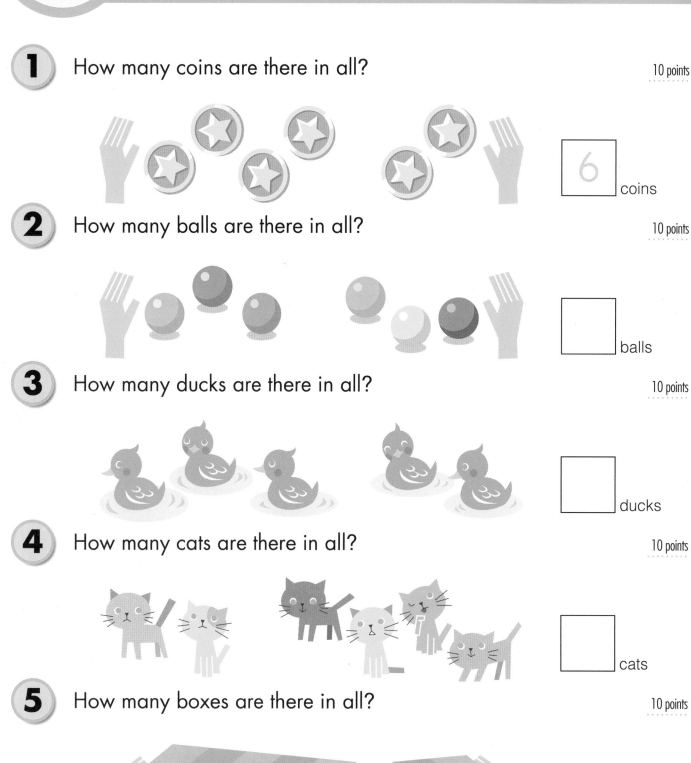

6 coins

2 How many balls are there in all?　　　　　10 points

balls

3 How many ducks are there in all?　　　　　10 points

ducks

4 How many cats are there in all?　　　　　10 points

cats

5 How many boxes are there in all?　　　　　10 points

boxes

　© Kumon Publishing Co., Ltd.

6 You have 3 boxes. Tom brings 4 more. How many total boxes do you have?

10 points

[ ] boxes

7 There are 4 cars in the parking lot. 2 more cars come and park. How many total cars are there in the parking lot?

10 points

[ ] cars

8 There are 5 birds eating food. 4 more birds come and eat. How many total birds are there?

10 points

[ ] birds

9 There are 3 chicks in a pen, and 6 chicks in another pen. How many total chicks are there in the pens?

10 points

[ ] chicks

10 Tom has 2 balloons and Sally has 4 balloons. How many balloons are there in all?

10 points

[ ] balloons

These are adding problems, right? Good job!

Addition

2

Date   /   /

Name

Level
★★

Score
_____ /100

---

1 If 3 apples and 2 apples are put together, there are 5 apples in all. Write this number sentence below.

10 points

$$3 + 2 = 5$$

2 Write number sentences about the pictures below.

10 points per question

(1) If **5** pencils and **3** pencils are put together, there are **8** pencils.

$$5 + 3 = 8$$

(2) If **2** eggs and **5** eggs are put together, there are **7** eggs.

$$\square + \square = \square$$

(3) If **4** fish and **2** fish are put together, there are **6** fish.

$$\square + \square = \square$$

(4) If **3** sheets of paper and **4** sheets of paper are put together, there are **7** sheets of paper.

$$\square + \square = \square$$

4   © Kumon Publishing Co., Ltd.

3 Write number sentences about the pictures below. 10 points per question

（1）

Altogether there are 7 buttons.

$\boxed{4} + \boxed{3} = \boxed{7}$

（2）

Altogether there are 8 children.

$\boxed{3} + \boxed{} = \boxed{}$

（3）

Altogether there are 6 birds.

$\boxed{} + \boxed{} = \boxed{}$

（4）

Altogether there are 9 dogs.

$\boxed{} + \boxed{} = \boxed{}$

（5）

Altogether there are 10 apples.

$\boxed{} + \boxed{} = \boxed{}$

"Altogether" means adding, too. Keep it up!

© Kumon Publishing Co., Ltd.

Date / /

Name

Level
★ ★

Score

/100

1 Write number sentences about the pictures below.

10 points per question

(1) How many birds are there altogether?

2 + 3 = ☐

⟨Ans.⟩ ☐ birds

(2) How many apples are there altogether?

3 + ☐ = ☐

⟨Ans.⟩ ☐ apples

(3) How many fish are there altogether?

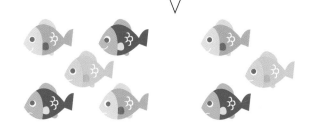

☐ + ☐ = ☐

⟨Ans.⟩ ☐ fish

(4) How many eggs are there altogether?

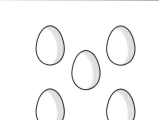 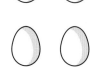

☐ + ☐ = ☐

⟨Ans.⟩ ☐ eggs

© Kumon Publishing Co., Ltd.

2 Write each number sentence below. Then answer the question.

10 points per question

(1) You have **5** sheets of blue paper and **4** sheets of white paper.
How many sheets of paper do you have?

$$5 + 4 = 9$$

⟨**Ans.**⟩ ☐ sheets

(2) There are **4** books on your desk, and you put down **2** more. How many books do you have?

$$4 + 2 =$$

⟨**Ans.**⟩ ☐ books

(3) There are **6** pencils in your pencil case, and **3** pencils on your desk. How many pencils do you have?

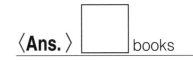

⟨**Ans.**⟩ ☐ pencils

(4) You have **3** eggs in the pan. You put in **2** eggs. How many eggs are in the pan?

⟨**Ans.**⟩ ☐ eggs

(5) There are **5** boys and **3** girls in line. How many children are in line?

⟨**Ans.**⟩ ☐ chidren

(6) **5** cars are parked. **2** more cars park. How many cars are parked?

⟨**Ans.**⟩ ☐ cars

Good job adding!

Addition

Date   /   /   Name

Level ★ ★

Score   /100

1 There are 3 red flowers and 6 yellow flowers. How many flowers are there altogether?   10 points

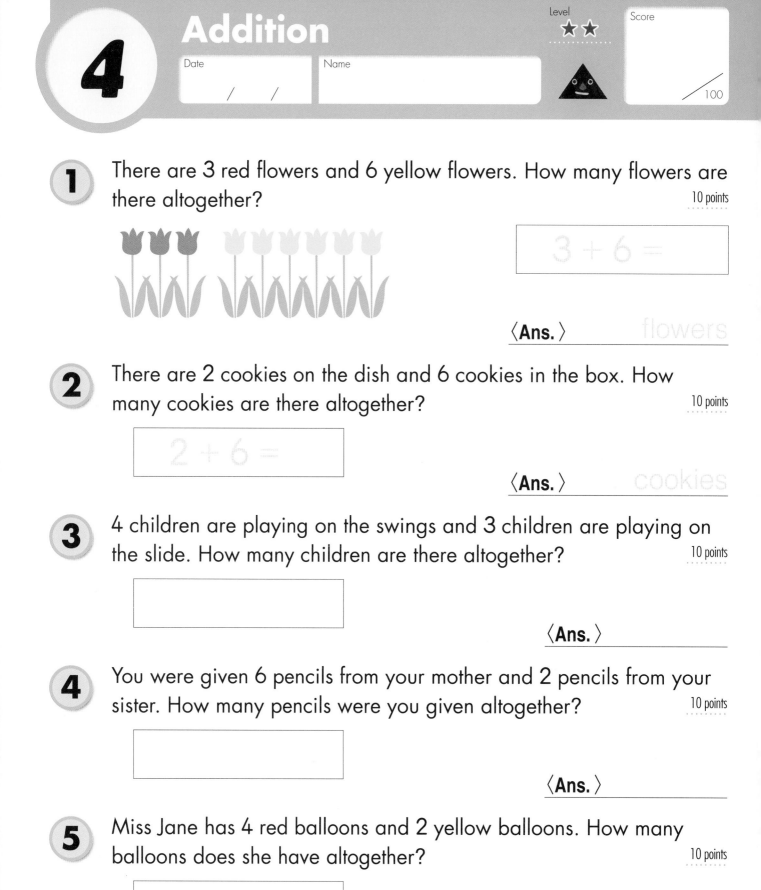

3 + 6 =

〈Ans.〉   flowers

2 There are 2 cookies on the dish and 6 cookies in the box. How many cookies are there altogether?   10 points

2 + 6 =

〈Ans.〉   cookies

3 4 children are playing on the swings and 3 children are playing on the slide. How many children are there altogether?   10 points

〈Ans.〉

4 You were given 6 pencils from your mother and 2 pencils from your sister. How many pencils were you given altogether?   10 points

〈Ans.〉

5 Miss Jane has 4 red balloons and 2 yellow balloons. How many balloons does she have altogether?   10 points

〈Ans.〉

© Kumon Publishing Co., Ltd.

6 4 flies land on the grass, and 5 flies fly in the air. How many flies are there altogether?

10 points

〈Ans.〉＿＿＿＿＿＿＿＿＿＿ flies

7 There are 3 red balls and 2 blue balls. How many balls are there altogether?

10 points

〈Ans.〉＿＿＿＿＿＿＿＿＿＿

8 There are 5 used pencils and 4 new pencils. How many pencils are there altogether?

10 points

〈Ans.〉＿＿＿＿＿＿＿＿＿＿

9 You picked 4 oranges. 3 oranges are left in the tree still. How many oranges are there altogether?

10 points

〈Ans.〉＿＿＿＿＿＿＿＿＿＿

10 6 white cars are parked. 2 black cars park, too. How many cars are there altogether?

10 points

〈Ans.〉＿＿＿＿＿＿＿＿＿＿

How many pencils do you have on your desk?
＿＿＿＿＿＿＿＿＿＿＿＿＿＿＿

Date   /   /

Name

1 There are 2 cars in the parking lot. If 3 more cars come, there are 5 cars altogether. Write the number sentence.

10 points

$$\boxed{2} + \boxed{3} = \boxed{}$$

2 If 2 more fish are put in the tank, there are 6 fish altogether.

10 points

$$\boxed{4} + \boxed{} = \boxed{}$$

3 Write each number sentence below.

10 points per question

(1)

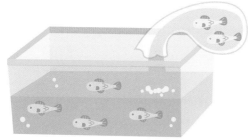

If Ted gives Jim 4 more pencils, Jim will have 7 pencils altogether.

$$\boxed{3} + \boxed{} = \boxed{}$$

(2)

If 2 more turtles come to the beach, there will be 5 turtles on the beach.

$$\boxed{} + \boxed{} = \boxed{}$$

© Kumon Publishing Co., Ltd.

4 Write each number sentence below.

10 points per question

(1) There are **3** flowers in the vase. You put **5** more flowers
in the vase, and now there are **8** flowers in all.

$$3 + 5 = 8$$

(2) **5** children were playing on the playground. After **2** more children came, there were **7**
children in all.

(3) There were **4** ducks in the pond, and **3** more were born. Now there are **7** ducks in all.

(4) You had **6** candies and your mother gave you **2** more. Now you have **8** candies in all.

(5) You had **7** books and your father gave you **2** more. Now you have **9** books in all.

(6) **4** sparrows were on the roof. **5** more sparrows landed on the roof, and now there are **9**
sparrows in all.

You are getting good!

© Kumon Publishing Co., Ltd.

Addition

1 Write each number sentence below. Then answer each question.

10 points per question

(1)

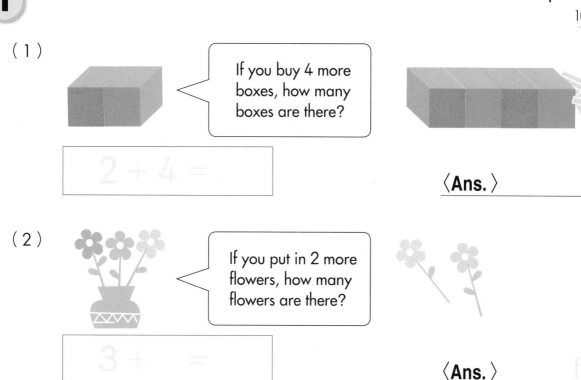

If you buy 4 more boxes, how many boxes are there?

2 + 4 =

⟨Ans. ⟩   boxes

(2)

If you put in 2 more flowers, how many flowers are there?

3 + =

⟨Ans. ⟩   flowers

(3)

If your mother gives you 3 more candies, how many candies will you have?

⟨Ans. ⟩

(4)

If 4 more turtles come to the beach, how many turtles will there be?

⟨Ans. ⟩

2 Write each number sentence below. Then answer each question.

10 points per question

(1) 4 cars are parked. 3 more cars park. How many cars are there in total?

4 + 3 =

〈Ans.〉 _____ cars

(2) 5 birds are eating food. 2 more birds come. How many birds are there in total?

〈Ans.〉 _____

(3) 6 children are in the playground. 2 more children come to play. How many children are there in total?

〈Ans.〉 _____

(4) You have 7 pencils. If I give you 2 pencils, how many total pencils will you have?

〈Ans.〉 _____

(5) There are 3 dogs eating. If 2 more dogs come and eat, how many total dogs will be eating?

〈Ans.〉 _____

(6) There are 4 apples on the table. If your mother puts 5 more apples on the table, how many total apples will be on the table?

〈Ans.〉 _____

Apples, cars and dogs—oh my!

1 6 children are jumping. If 2 more children start jumping, how many children will be jumping?

10 points

⟨Ans.⟩

2 There are 4 fish in the pond. If you put in 3 more fish, how many fish will there be?

10 points

⟨Ans.⟩

3 You have 7 sheets of colored paper. If your mother gives you 2 more sheets of colored paper, how many sheets of colored paper will you have?

10 points

⟨Ans.⟩

4 You saw 5 fish in the pond. Then you saw 3 more. How many fish did you see?

10 points

⟨Ans.⟩

5 You have 6 comic books. If you buy 3 more comic books, how many comic books will you have?

10 points

⟨Ans.⟩

© Kumon Publishing Co., Ltd.

6 Pat cooked 4 hot dogs. Then he cooked 3 more. How many hot dogs did he cook?

10 points

⟨Ans. ⟩ _____

7 The hens laid 5 eggs. Then they laid 4 more. How many eggs did the hens lay?

10 points

⟨Ans. ⟩ _____

8 You made 4 paper toys. Then you made 2 more. How many paper toys did you make?

10 points

⟨Ans. ⟩ _____

9 Emmy ate 3 oranges. Then she ate 2 more. How many oranges did Emmy eat?

10 points

⟨Ans. ⟩ _____

10 6 sparrows were in the garden. Then 3 more came. How many sparrows were there in all?

10 points

⟨Ans. ⟩ _____

Is it getting easier? Good!

Addition

8

Date   /   /

Name

Level
★ ★ ★

Score

/100

1 There were 5 fish swimming in the pond. You put 4 more into the pond. How many fish are in the pond now?

10 points

⟨Ans.⟩   fish

2 Ken has 4 storybooks and 6 picture books. How many books does he have altogether?

10 points

⟨Ans.⟩

3 There are 6 sheets of red paper and 3 sheets of blue paper. How many sheets of paper are there altogether?

10 points

⟨Ans.⟩

4 There were 6 birds sitting on the tree. Then 2 birds landed. How many birds are on the tree now?

10 points

⟨Ans.⟩

5 George's family ate 3 eggs yesterday and 4 eggs today. How many eggs did they eat in all?

10 points

⟨Ans.⟩

6 9 children were playing in the park. Then 3 more children came to the park. How many children were at the park altogether? 10 points

⟨Ans.⟩

7 Sue had 7 books. Then she bought 3 books. How many books does she have in all? 10 points

⟨Ans.⟩

8 2 ships were in the harbor. Then 6 ships came back to the harbor. How many ships were there altogether? 10 points

⟨Ans.⟩

9 You want to send 8 cards to your friends and 4 cards to your family. How many cards do you need? 10 points

⟨Ans.⟩

10 Sam gave John 3 pencils. Sam also gave Kate 4 pencils. How many pencils did Sam give away? 10 points

⟨Ans.⟩

Are you ready for something a little different?

1 There are 5 oranges. The number of pears is 1 more than the number of oranges. How many pears are there?

10 points

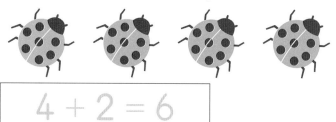

⟨Ans.⟩ 6 pears

2 There are 5 storybooks on the shelf. There are 2 more picture books than storybooks. How many picture books are there?

10 points

⟨Ans.⟩

3 There are 4 ladybugs in the garden. There are 2 more butterflies than ladybugs. How many butterflies are there?

10 points

4 + 2 = 6

⟨Ans.⟩

4 There are 6 canaries. There are 4 more chickens than canaries. How many chickens are there?

10 points

⟨Ans.⟩

© Kumon Publishing Co., Ltd.

5 There are 5 boys watching the movie. There are 3 more girls than boys watching the movie. How many girls are there?

10 points

〈Ans.〉 _____

6 There are 3 sheets of red paper on the desk. There are 5 more sheets of blue paper than red paper. How many sheets of blue paper are there?

10 points

〈Ans.〉 _____

7 There are 4 empty flowerpots. There are 2 more bulbs than flowerpots. How many bulbs?

10 points

〈Ans.〉 _____

8 Diana got 5 oranges from her mother. Jack got 4 more oranges than Diana did. How many oranges did Jack get?

15 points

〈Ans.〉 _____

9 Bob ate 7 strawberries from the bowl. Betty ate 3 more strawberries than Bob. How many strawberries did Betty eat?

15 points

〈Ans.〉 _____

How is it going? Good I hope!

Addition

Date / /　　Name

1 You must put one cookie on each dish. There are 5 dishes. How many cookies do you need?

10 points

⟨Ans.⟩ 5 cookies

2 You must give one sheet of paper to each child. There are 2 boys and 3 girls. How many sheets of paper do you need?

10 points

⟨Ans.⟩ _____

3 You must give one pencil to each child. There are 3 boys and 4 girls in the class. How many pencils do you need?

10 points

3 + 4 = 7

⟨Ans.⟩ _____

4 There were 4 cats playing in the yard, and then 2 more cats came. If you want to put 1 bell on each cat, how many bells do you need?

10 points

⟨Ans.⟩ _____

5 There are 3 red flowers and 5 white flowers at the store. If you want to put one flower in each vase, how many vases do you need?   10 points

⟨**Ans.**⟩ _____

6 You want to send 3 boys and 6 girls one card each. How many cards do you have to send?   10 points

⟨**Ans.**⟩ _____

7 There are 4 apples and 5 pears in the kitchen. If you want to pack one fruit in each lunchbag, how many bags will you need?   10 points

⟨**Ans.**⟩ _____

8 4 red ships and 4 blue ships land at the harbor. If you want to put a flag on each ship, how many flags would you need?   15 points

⟨**Ans.**⟩ _____

9 6 children are sitting on chairs. 4 chairs aren't being used. How many chairs are there in all?   15 points

⟨**Ans.**⟩ _____

Getting used to the new questions?
Well done.

Addition

1 6 dogs were playing in the park. 2 more dogs came. How many dogs were in the park?

10 points

⟨Ans.⟩ _____

2 You ate 8 chocolate chip cookies yesterday. Today, you ate 4 sugar cookies. How many cookies did you eat altogether?

10 points

⟨Ans.⟩ _____

3 There are 7 fish in the pond. There are 4 fish in the tank. How many fish are there in all?

10 points

⟨Ans.⟩ _____

4 5 children are playing in the sandbox. 4 children are on the slide. How many total children are playing?

10 points

⟨Ans.⟩ _____

5 You gave 3 children an orange each, and then you also gave 5 children an orange each. How many oranges did you give the children?

10 points

⟨Ans.⟩ _____

6 6 cars were racing. 2 more cars joined the race. How many cars were racing?

10 points

⟨Ans.⟩ _____

7 There are 8 comic books. There are 2 more picture books than comic books. How many picture books are there?

10 points

⟨Ans.⟩ _____

8 Mary made 3 paper airplanes. Tom made 5 more paper airplanes than Mary. How many paper airplanes did Tom make?

10 points

⟨Ans.⟩ _____

9 There were 8 children in class. Then 4 children came late. How many children are in class altogether?

10 points

⟨Ans.⟩ _____

10 There are 6 apples in the kitchen. There are 3 more oranges than apples in the kitchen. How many oranges are in the kitchen?

10 points

⟨Ans.⟩ _____

You're really starting to get it! Way to go!

Subtraction

1 If you take away 2 buttons, how many remain?   10 points

 buttons
6

2 If you take away 3 boxes, how many remain?   10 points

 boxes

3 If 4 sparrows fly away, how many remain?   10 points

 sparrow(s)

4 If you eat 3 strawberries, how many remain?   10 points

 strawberries

5 If the boy catches 6 fish, how many remain?   10 points

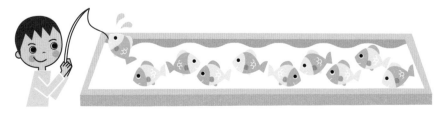

 fish

© Kumon Publishing Co., Ltd.

6 How many more apples are there than oranges?

10 points

apples

7 How many more girls are there than boys?

10 points

Boys

Girls

girls

8 How many more dogs are there than cats?

10 points

dogs

9 There were 5 frogs on the lily pad. Then 2 of them jumped into the water for a swim. How many frogs are left on the lily pad?

10 points

frogs

10 How many more textbooks than comic books are there?

10 points

textbooks

We are working with subtraction now.
Good job!

© Kumon Publishing Co., Ltd.   25

Subtraction

13

Date / /

Name

Level ★★

Score /100

1 There are three boxes on the table. After you take 1 box away, 2 boxes remain. Write the number sentence.

10 points

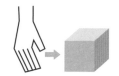

$3 - 1 = 2$

2 Write each number sentence below.

10 points per question

(1) There were **5** turtles in the pond. After **2** turtles got out of the pond, **3** turtles remained in the pond.

$5 - 2 = 3$

(2) There were **6** sheets of paper in the folder. After you used **2** sheets of paper to make paper rings, **4** sheets remained in the folder.

$6 - 2 = 4$

(3) There were **7** birds on the line. After **3** birds flew away, **4** birds remained on the line.

$\square - \square = \square$

(4) You had **8** flowers in your bunch. After you gave **2** flowers to a friend, **6** flowers are left in your bunch.

$\square - \square = \square$

3 Write each number sentence below.

10 points per question

(1)

If 4 birds fly away, 2 birds remain.

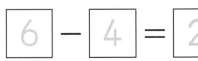

$$6 - 4 = 2$$

(2)

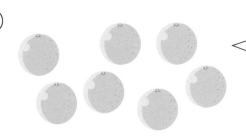

If you eat 2 oranges, there are 5 oranges remaining.

$$7 - \square = \square$$

(3)

If you break 3 pencils, there are 5 pencils remaining.

$$\square - \square = \square$$

(4)

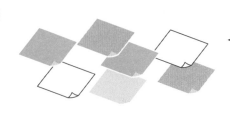

If you use 5 sheets of paper, there are 2 sheets left.

$$\square - \square = \square$$

(5)

If 4 children go home, there are 5 children still at the sandbox.

$$\square - \square = \square$$

See how these sentences all mean you should take away? You're smart!

Subtraction

Date ____ / ____ / ____

Name _____

Level

Score _____ /100

---

1 Write each number sentence below. Then answer each question.

10 points per question

(1)

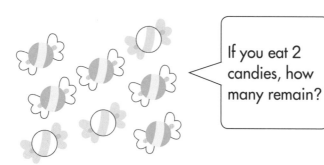

If you eat 2 candies, how many remain?

$\boxed{8} - \boxed{2} = \boxed{}$

⟨**Ans.**⟩ $\boxed{}$ candies

(2)

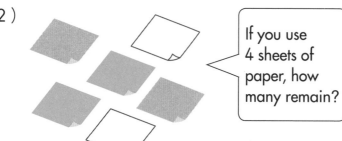

If you use 4 sheets of paper, how many remain?

$\boxed{6} - \boxed{} = \boxed{}$

⟨**Ans.**⟩ $\boxed{}$ sheets

(3)

If you eat 3 pieces of cake, how many are left?

$\boxed{} - \boxed{} = \boxed{}$

⟨**Ans.**⟩ $\boxed{}$ pieces

(4)

If you give 3 fish to your sister, how many do you have left?

$\boxed{} - \boxed{} = \boxed{}$

⟨**Ans.**⟩ $\boxed{}$ fish

 © Kumon Publishing Co., Ltd.

2 Write each number sentence below. Then answer each question.

10 points per question

(1) There were **6** sparrows on the electric wire.
After **2** sparrows flew away, how many were left?

$$6 - 2 = 4$$

⟨**Ans.**⟩ ☐ sparrows

(2) There are **10** pencils on your desk. If you give **3**
pencils away, how many pencils will remain on your
desk?

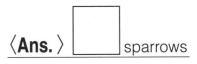

$$10 - 3 =$$

⟨**Ans.**⟩ ☐ pencils

(3) There are **8** fish in the pond. If you take **5** fish away, how many fish are left in the pond?

⟨**Ans.**⟩ ☐ fish

(4) There were **7** sheets of paper. If you used **4** sheets of paper for coloring, how many
sheets of paper do you have left?

⟨**Ans.**⟩ ☐ sheets

(5) There are **9** eggs. If you eat **3** eggs, how many eggs will you have left to eat?

⟨**Ans.**⟩ ☐ eggs

(6) There are **8** flowers in the garden. If you take **6** away, how many will remain?

⟨**Ans.**⟩ ☐ flowers

Are you getting the hang of it?
Good!

Subtraction

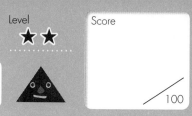

1 There are 8 apples. If we eat 5 apples, how many remain?

10 points

⟨Ans.⟩

2 9 cars are in the parking lot. If 3 cars go home, how many cars are left in the parking lot?

10 points

⟨Ans.⟩

3 You have 7 dollars. If you use 4 dollars to buy lunch, how many dollars will you have left?

10 points

⟨Ans.⟩

4 8 children are playing in the park. If 3 children go home, how many remain?

10 points

⟨Ans.⟩

5 I had 7 balloons. After 3 balloons flew away, how many balloons did I have left?

10 points

⟨Ans.⟩

  © Kumon Publishing Co., Ltd.

6 You had 6 picture books. After you lent 2 picture books to your friend, how many did you have left? 10 points

⟨Ans.⟩ _____

7 There were 8 pears on the tree. After you took 5 pears away, how many pears remained on the tree? 10 points

⟨Ans.⟩ _____

8 Grandma had 7 oranges. After Jonny ate 2 oranges, how many did Grandma still have? 10 points

⟨Ans.⟩ _____

9 Ken had 9 candles. Last night, he used 5 candles when the power went out. How many candles does Ken still have? 10 points

⟨Ans.⟩ _____

10 There were 10 fish in the pond. After you gave 2 fish to your friend, how many fish remained in the pond? 10 points

⟨Ans.⟩ _____

No problem, right? You are a star!

© Kumon Publishing Co., Ltd.

16

Subtraction

Level
★ ★

Date
/ /

Name

Score
/100

1 I have 7 oranges. If I eat 3 oranges, how many do I have left? 10 points

⟨Ans.⟩

2 Tom had 6 sheets of paper. After he used 2 sheets of paper, how many sheets remained? 10 points

⟨Ans.⟩

3 The teacher has 7 pencils. 3 pencils have been sharpened. How many pencils are not sharp? 10 points

7 − 3 =

⟨Ans.⟩

4 You got 6 picture books for your birthday. You have read 2 so far. How many picture books have you not read? 10 points

⟨Ans.⟩

© Kumon Publishing Co., Ltd.

5 You have 9 dollars. You used 4 dollars to buy a snack. How many dollars do you have left?

10 points

⟨Ans.⟩ _____

6 There are 6 tomatoes for dinner. 2 of them are bad. How many tomatoes are still good?

10 points

⟨Ans.⟩ _____

7 Jack threw 10 balls. 6 balls went into the basket. How many balls did not go in the basket?

10 points

⟨Ans.⟩ _____

8 There are 7 umbrellas in the hall. 3 of them are broken. How many umbrellas still work?

10 points

⟨Ans.⟩ _____

9 There are 8 total red and yellow balloons. There are 3 red balloons. How many yellow balloons are there?

10 points

⟨Ans.⟩ _____

10 8 children are playing tag. 5 of them are boys. How many girls are playing tag?

10 points

⟨Ans.⟩ _____

Good job!

1 How many more cats are there than mice?   10 points

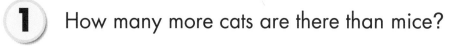

⟨Ans.⟩   2   cats

2 There are 7 desks and 5 chairs in the room. How many more desks than chairs are there?   10 points

7 − 5 = 2

⟨Ans.⟩

3 There are 8 bicycles and 6 unicycles. How many more bicycles are there than unicycles?   10 points

⟨Ans.⟩

4 I have 9 red balloons and 5 blue balloons. How many more red balloons do I have than blue balloons?   10 points

⟨Ans.⟩

5 How many more oranges than apples are there?   10 points

⟨Ans.⟩ _____

6 How many more marbles are there on the right than on the left?

10 points

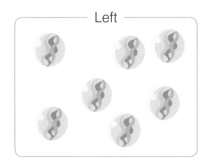   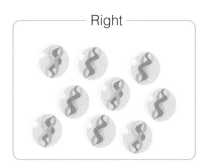

⟨Ans.⟩ _____

7 There are 8 frogs and 3 fish in the pond. How many more frogs than fish are there in the pond?   20 points

⟨Ans.⟩ _____

8 My garden has 4 red flowers and 7 white flowers. How many more white flowers than red flowers do I have?   20 points

⟨Ans.⟩ _____

Now you're moving. Way to go!

Date   /   /

Name

1 How many more apples than oranges are there?

10 points

⟨Ans.⟩ 3 apples

2 How many fewer oranges than apples are there?

10 points

⟨Ans.⟩

3 There are 6 apples and 4 oranges. How many fewer oranges than apples are there?

10 points

6 − 4 = 2

⟨Ans.⟩

4 There are 6 butterflies and 2 ladybugs. How many fewer ladybugs are there?

10 points

⟨Ans.⟩

5 How many more monkeys than bears are there?   15 points

8 − 5 =

⟨Ans.⟩ _____

6 How many fewer dishes than cookies are there?   15 points

⟨Ans.⟩ _____

7 Our class has 9 boys and 6 girls. How many more boys than girls does our class have?

15 points

⟨Ans.⟩ _____

8 You have 4 marbles and 10 blocks. How many fewer marbles than blocks do you have?

15 points

⟨Ans.⟩ _____

Are you getting it? Good!

© Kumon Publishing Co., Ltd.   37

Subtraction

Date   /   /   Name

Level ★★★

Score   /100

---

1 There are 8 apples and 6 oranges. How many more apples than oranges are there?

10 points

⟨**Ans.**⟩ _____

2 There are 4 apples and 7 oranges. How many fewer apples than oranges are there?

10 points

⟨**Ans.**⟩ _____

3 You have 8 apples and 5 oranges. Which fruit do you have more of? How many more?

10 points

⟨**Ans.**⟩ I have ☐ more | apples |.

4 You have 9 picture books and 7 storybooks. Which kind of book do you have more of? How many more?

10 points

⟨**Ans.**⟩ I have _____ more _____.

  © Kumon Publishing Co., Ltd.

5 There are 4 butterflies and 6 dragonflies. Are there more dragonflies or butterflies? How many more?

15 points

⟨**Ans.**⟩ There are _____ more _____ .

6 There are 5 red pencils and 3 blue pencils. Which color pencil do you have fewer of? How many fewer?

15 points

⟨**Ans.**⟩ There are [ ] fewer blue pencils .

7 There are 8 flower bulbs and 5 pots. Are there fewer pots or bulbs? How many fewer are there?

15 points

⟨**Ans.**⟩ There are _____ fewer _____ .

8 There are 7 hats and 10 hat hooks. Are there fewer hats or hat hooks? How many fewer are there?

15 points

⟨**Ans.**⟩ _____

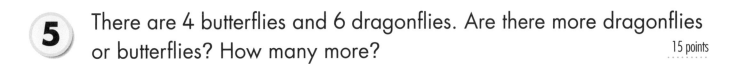

Do you have more apples or oranges at your house? Go check!

20 Subtraction

Level ★★

Date / /

Name

Score

/100

1 I have 8 apples. I have 3 fewer oranges than apples. How many oranges do I have?

10 points

8 − 3 = 5

⟨Ans.⟩

2 Our class has 6 boys. We have 3 fewer girls than boys. How many girls are there?

10 points

6 − =

⟨Ans.⟩

3 7 bicycles are at the park. There are 3 fewer unicycles at the park. How many unicycles are there at the park?

10 points

⟨Ans.⟩

4 Your grandmother has 8 bulbs. She has 2 fewer pots. How many pots does your grandmother have?

10 points

⟨Ans.⟩

 © Kumon Publishing Co., Ltd.

5 The clown has 10 red balloons. He has 2 fewer yellow balloons than red balloons. How many yellow balloons does he have?

10 points

⟨Ans.⟩ _____

6 I have 9 crayons. I have 3 fewer pencils than crayons. How many pencils do I have?

10 points

⟨Ans.⟩ _____

7 Jack got 8 peaches for lunch. Sue got 5 fewer peaches than Jack. How many peaches did Sue get?

10 points

⟨Ans.⟩ _____

8 There are 7 white hair clips in the bathroom. There are 2 fewer red clips than white clips. How many red clips are in the bathroom?

10 points

⟨Ans.⟩ _____

9 Mike has 9 stickers. Jim has 4 fewer stickers than Mike. How many stickers does Jim have?

10 points

⟨Ans.⟩ _____

10 The hens laid 6 eggs yesterday. Today they laid 1 fewer egg. How many eggs did the hens lay today?

10 points

⟨Ans.⟩ _____

Subtraction is no problem, right?

21 Subtraction

Level
★ ★

Date
/ /

Name

Score
/100

1 There are 5 chairs. If 3 children sit on a chair, how many chairs are left?

10 points

⟨Ans.⟩ 2 chairs

2 The teacher has 6 apples. If she gives 1 apple each to 4 children, how many apples will remain?

10 points

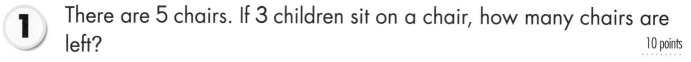

⟨Ans.⟩ _____

3 There are 6 marbles. If 3 children get 1 marble each, how many marbles will be left?

10 points

6 − 3 = 3

⟨Ans.⟩ _____

4 Grandmother baked 7 cookies. If she gives 1 cookie each to 5 children, how many cookies will be left?

10 points

⟨Ans.⟩ _____

5 In the kitchen, there are 8 cakes and 5 dishes. If you put 1 cake on each dish, how many more dishes will you need?

10 points

⟨Ans.⟩ _____

6 You have 8 bells. If you put 1 bell each on 6 cats, how many bells will be left?

10 points

⟨Ans.⟩ _____

7 In class today, there are 6 chairs and 8 people. If everyone sits down, how many more chairs will you need?

10 points

⟨Ans.⟩ _____

8 I gave 1 orange each to 8 children. I had 10 oranges. How many oranges do I have now?

10 points

⟨Ans.⟩ _____

9 The teacher gave 1 sheet of paper each to 6 children. There were 9 sheets of paper. How many sheets of paper remain?

10 points

⟨Ans.⟩ _____

10 At lunch, there are 7 children. The teacher wants to give 1 bottle of milk to each child, but there are only 5 bottles of milk. How many more bottles of milk does the teacher need?

10 points

⟨Ans.⟩ _____

Don't worry if it's hard. Just keep trying!

22

Subtraction

Date   /   /

Name

Level ★ ★

Score
/100

1 Yesterday, you had 5 apples. Today, you ate 2 of them. How many apples remain?

10 points

⟨Ans.⟩ _____

2 Jim had 8 pencils, but he broke 3 of them in class. How many pencils does he have left?

10 points

⟨Ans.⟩ _____

3 Sarah found 6 shells on the beach. She lost 4 on the way home. How many shells did she have left when she got home?

10 points

⟨Ans.⟩ _____

4 In Tom's garden, there are 7 red flowers and 3 white flowers. How many more red flowers than white flowers does Tom have?

10 points

⟨Ans.⟩ _____

5 Ann used 9 stickers today. Henry used 3 fewer stickers. How many stickers did Henry use?

10 points

⟨Ans.⟩ _____

6 8 children were playing in the park. 2 children went home. How many children are still playing?

10 points

⟨Ans.⟩ _____

7 7 boys and 4 girls are playing tag. Are there more boys or girls, and how many more are there?

10 points

⟨Ans.⟩ There are _____ more _____ .

8 Kate and Jim are jumping rope. She jumped 8 times and he jumped 10 times. Who jumped more times, and how many more times did they jump?

10 points

⟨Ans.⟩ _____ jumped _____ more times.

9 Kate and Jim are jumping rope. She jumped 8 times and he jumped 10 times. What is the difference in the number of times they jumped?

10 points

⟨Ans.⟩ _____

10 There are 9 fish altogether in a tank full of red and black fish. There are 6 red fish. How many black fish are there?

10 points

⟨Ans.⟩ _____

Okay—now we are going to mix it up!

1 You had 8 sheets of paper, and then you used 2 sheets for homework. How many blank sheets of paper do you have left? 10 points

⟨Ans.⟩ _____

2 You had 8 sheets of paper, and then you received 2 sheets from your mother. How many sheets of paper do you have altogether?

10 points

⟨Ans.⟩ _____

3 Jimmy is 6 years old. His brother is 3 years older than him. How old is his brother? 10 points

⟨Ans.⟩ _____

4 Timmy is 6 years old, too. His brother is 3 years younger than him. How old is his younger brother? 10 points

⟨Ans.⟩ _____

5 There are 5 big fish and 4 small fish in my tank. How many fish are there altogether? 10 points

⟨Ans.⟩ _____

© Kumon Publishing Co., Ltd.

6 For the picnic, Mother got 6 watermelons and 8 peaches. How many more peaches are there?

10 points

〈Ans.〉 _____

7 Polly found 6 shells at the beach. Rob found 4 more than her. How many shells did Rob find?

10 points

〈Ans.〉 _____

8 There are 9 cows and 7 sheep at the farm. How many more cows than sheep are there?

10 points

〈Ans.〉 _____

9 The milkman has 7 bottles. 2 of them are empty. How many full bottles of milk does the milkman have?

10 points

〈Ans.〉 _____

10 Fred ran around the pond 9 times. Kim did it 6 times. Who ran around the pond more times, and how many more times did he or she do it?

10 points

〈Ans.〉 _____

Do you know when to add or subtract? Good!

Addition or Subtraction

Level ★★★

1 There were 7 eggs in the kitchen. Today, your family ate 5 eggs. How many eggs remain?

10 points

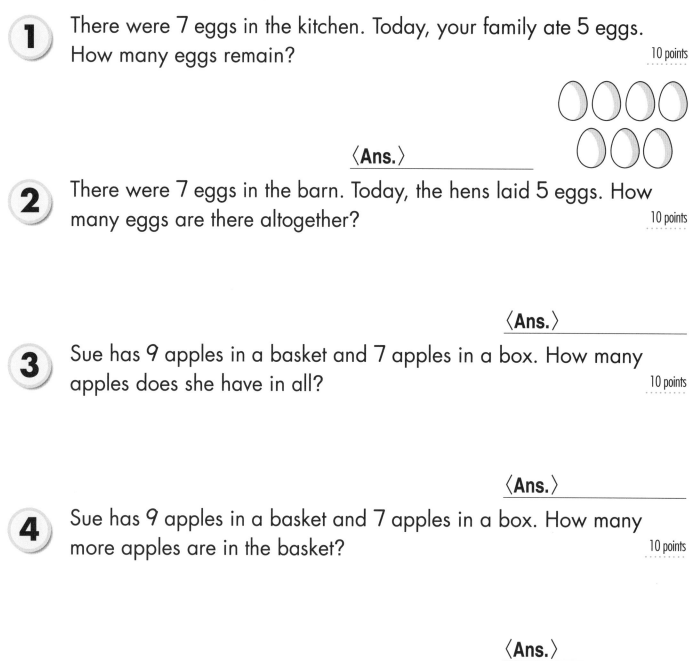

⟨Ans.⟩ _____

2 There were 7 eggs in the barn. Today, the hens laid 5 eggs. How many eggs are there altogether?

10 points

⟨Ans.⟩ _____

3 Sue has 9 apples in a basket and 7 apples in a box. How many apples does she have in all?

10 points

⟨Ans.⟩ _____

4 Sue has 9 apples in a basket and 7 apples in a box. How many more apples are in the basket?

10 points

⟨Ans.⟩ _____

5 Grandmother has 16 grandchildren. 9 of them are boys. How many girls are there?

10 points

⟨Ans.⟩ _____

© Kumon Publishing Co., Ltd.

6 Jane had 9 stickers, and then she got 4 more from her mother. How many stickers does she have?

10 points

⟨Ans.⟩ _____

7 Jane has 9 stickers. Susie has 4 more stickers than Jane. How many stickers does Susie have?

10 points

⟨Ans.⟩ _____

8 The pond has 6 frogs. There are 5 more goldfish than frogs. How many goldfish are there in the pond?

10 points

⟨Ans.⟩ _____

9 On the farm, there are 8 horses and 15 cows. How many more cows than horses are there?

10 points

⟨Ans.⟩ _____

10 14 children have to get their flu shots. 9 are already finished. How many children still need to get their flu shots?

10 points

⟨Ans.⟩ _____

This is tough, but you can do it!

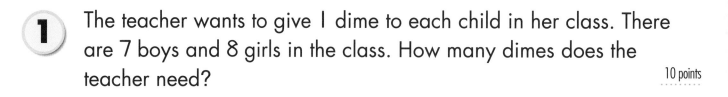
1 The teacher wants to give 1 dime to each child in her class. There are 7 boys and 8 girls in the class. How many dimes does the teacher need?

10 points

〈Ans.〉

2 Tom has 8 pencils. If he gives 1 pencil to each child, and there are 13 children in the class, how many children won't get pencils?

10 points

〈Ans.〉

3 In the parking lot, there are 9 bikes. There are 6 more cars than bikes. How many cars are in the parking lot?

10 points

〈Ans.〉

4 There are 9 bikes and 6 cars in the lot today. Are there more bikes or cars? How many more?

10 points

〈Ans.〉 There are _____ more _____ .

5 There are 14 blue stickers altogether. There are 6 fewer red stickers than blue stickers. How many red stickers are there?

10 points

〈Ans.〉

6 Mary had 6 pennies. Her sister gave her 8 more. How many pennies does Mary have now?

10 points

〈Ans.〉 _____

7 You had 15 flowers, and then you gave your friend 9 of them. How many flowers do you have left?

10 points

〈Ans.〉 _____

8 Jamal had 14 toy cars, but then his little brother broke 6 of them. How many cars does Jamal have now that are not broken?

10 points

〈Ans.〉 _____

9 John got 7 chestnuts and Debby got 8 chestnuts. How many chestnuts are there in all?

10 points

〈Ans.〉 _____

10 I have 15 stamps. If I use 7 stamps, how many stamps remain?

10 points

〈Ans.〉 _____

These are hard. Good job!

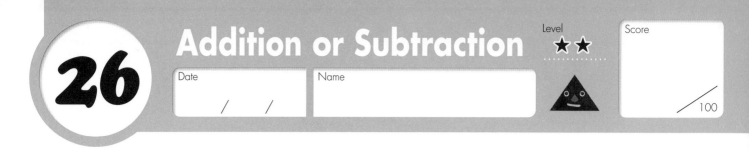

1 A bus carrying 6 people stopped at a bus stop. 4 people got on the bus. How many people are now on the bus in all?

10 points

There were 6 people on the bus.

⟨**Ans.**⟩

2 A bus carrying 8 people stopped at a bus stop, and 5 people got out. How many people are now on the bus?

10 points

There were 8 people on the bus.

⟨**Ans.**⟩

3 There were 6 people on the bus. At the first stop, 4 people got on the bus. At the next stop, 3 more people got on the bus. How many people are now on the bus?

20 points

There were 6 people on the bus.

⟨**Ans.**⟩ 13 people

4 There were 6 people on the bus. 4 people got on the bus at a bus stop. At the next stop, 3 people got off the bus. How many people are now on the bus?

20 points

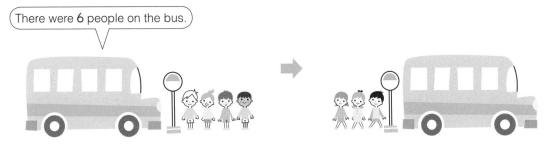

There were 6 people on the bus.

〈Ans.〉 _____

5 Jane had 5 tokens for the games. Her mother gave her 3 more tokens, and her sister gave her 2 more tokens. How many tokens does she have in all?

20 points

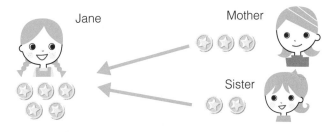

Jane   Mother

Sister

〈Ans.〉 _____

6 Kate had 4 stickers. She was given 6 stickers from her sister. Then she used 3 stickers. How many stickers does she have now?

20 points

〈Ans.〉 _____

Just go one number at a time, and you will have no problem!

1 You had 7 candies. You were given 3 candies from your mother and 2 candies from your sister. How many candies do you have now? 10 points

$$\boxed{7} + \boxed{3} + \boxed{2} = \boxed{12}$$

⟨Ans.⟩

2 You had 7 candies. You were given 3 candies from your mother, and then you gave your sister 2 candies. How many candies do you have now? 10 points

$$\boxed{7} + \boxed{3} - \boxed{2} = \boxed{8}$$

⟨Ans.⟩

3 You had 7 candies. You gave your sister 3 candies, and then you were given 2 candies from your mother. How many candies do you have now? 10 points

$$\boxed{7} - \boxed{3} + \boxed{2} = \boxed{}$$

⟨Ans.⟩

4 You had 7 candies. You gave your sister 3 candies and your brother 2 candies. How many candies do you have now? 10 points

$$\boxed{7} - \boxed{} - \boxed{} = \boxed{}$$

⟨Ans.⟩

5 Helen had 9 nickels. She used 3 nickels, and then she gave her sister 4 nickels. How many nickels does she have now? 10 points

$$9 - 3 - 4 =$$

⟨Ans.⟩

6 Ken had 4 stickers. He used 2 of them, and then his mother gave him 8. How many stickers does Ken have now? 10 points

$$4 - 2 + 8 =$$

⟨Ans.⟩ _____

7 You had 5 apples. Mrs. Brown gave you 3 apples. Then you used 4 apples in a pie. How many apples do you have now? 10 points

$$5 + \quad - \quad =$$

⟨Ans.⟩ _____

8 4 people were on a bus. 6 people got on the bus at a stop. At the next stop, 3 more people got on. How many people are on the bus now? 10 points

$$4 +$$

⟨Ans.⟩ _____

9 Tim found a pond with 9 fish in it. He got 2 fish from it one day. Yesterday, he got 3 fish. How many fish are left in the pond today? 10 points

$$9 -$$

⟨Ans.⟩ _____

10 Ann got 5 hair clips yesterday. Today, she broke 3 hair clips, and her uncle gave her 8 new ones. How many hair clips does she have now? 10 points

$$5 -$$

⟨Ans.⟩ _____

No problem, right?

1 You had 10 sheets of paper. You used 4 for paper airplanes. Then you gave your friends 3 sheets of paper. How many sheets do you have left? 10 points

$$10 - 4 - 3 =$$

⟨Ans.⟩

2 8 children were playing at the park. Then 2 children came to the park, and 3 children went home. How many children are playing at the park now? 10 points

⟨Ans.⟩

3 Sophie had 4 marbles. Then her sister gave Sophie 6 marbles, and her brother took 3 marbles away. How many marbles does Sophie have now? 10 points

⟨Ans.⟩

4 5 birds were eating some food. 2 birds landed, and then 2 more birds landed. How many birds are there now? 10 points

⟨Ans.⟩

5 There are 8 people on a bus. 4 people exit, and then 3 people get on the bus. How many people are on the bus now? 10 points

⟨Ans.⟩

6 9 dogs were sleeping. 3 dogs woke up and went outside, and then 4 more left. How many dogs are still sleeping?

10 points

⟨Ans.⟩ _____

7 Bob had 4 seeds. He planted 3 of them, but then his father gave him 5 more. How many seeds does Bob have now?

10 points

⟨Ans.⟩ _____

8 6 ducks were floating around a pond. 2 ducks got out of the pond, and 5 ducks got into the pond. How many ducks are in the pond?

10 points

⟨Ans.⟩ _____

9 John was shooting his basketball. At first, he made 4 baskets. The second time, he made 3 baskets, so he tried again and made 3 more. How many baskets did John make in all?

10 points

⟨Ans.⟩ _____

10 You had 5 stamps. You used 4 stamps, and then you bought 7 more. How many stamps do you have now?

10 points

⟨Ans.⟩ _____

Okay, let's try something different!

Ordinal Numbers

29

Date   /   /

Name

Level
★ ★

Score

/100

---

1 Answer the questions below.

10 points per question

(1) Circle **6** children from the front.

(2) Circle the sixth child from the front.

(3) Circle the child that is third behind John.

John

(4) What number from the front is the person that is third behind John? Take a long look at the picture before answering.

〈Ans.〉 _____

2 George is third from the front. How far from the front is the second child past George?

10 points

George

3 + 2 = 5

〈Ans.〉 Fifth

58   © Kumon Publishing Co., Ltd.

3 Roy is fifth from the front of the line. Bob is third behind Roy. What number child from the front is Bob?

10 points

⟨Ans.⟩ _____

4 Children form a row in gym class. Meg is fourth from the left, and Julie is fifth from Meg. What number child from the left is Julie?

10 points

⟨Ans.⟩ _____

5 A story book is fifth from the right in a bookshelf. An animal book is two books past the story book from the right. What number from the right is the animal book?

15 points

⟨Ans.⟩ _____

6 Sam and Beth are on the stairs. Sam is on the fourth step from the bottom. Beth is on the fifth step past Sam. What number from the bottom is Beth?

15 points

⟨Ans.⟩ _____

You're starting to get this, right?

© Kumon Publishing Co., Ltd. 59

Ordinal Numbers

1 The class is waiting in a line for lunch. There are 5 children in front of Cindy in line.

10 points per question

(1) Circle Cindy in the picture.

(2) What number child from the front is Cindy?

$$\boxed{5} + 1 = \boxed{6}$$

⟨Ans.⟩ Sixth

2 5 children finish in front of David in a race. What place is David in?

10 points

$$5 + \boxed{1} = \boxed{6}$$

⟨Ans.⟩ _____

3 Everyone is waiting in line for the bus. There are 6 children in front of Ellen. What number child from the front is Ellen?

10 points

$$6 + 1 =$$

⟨Ans.⟩ _____

4 There is a comic book on the bookshelf. There are 8 books between the comic book and the left side of the shelf. What number book from the left is the comic book?

10 points

⟨Ans.⟩ _____

   © Kumon Publishing Co., Ltd.

5 Today everyone has to get their shots. 7 children get their shots and Fred is next. What number child from the front was Fred? 10 points

⟨Ans.⟩ _____

6 A line of cars is waiting for the toll. Andy is in the sixth car from the front. How many cars are in front of his car? 10 points

$\boxed{6} - 1 = \boxed{5}$

⟨Ans.⟩ _____

7 A line of children is waiting to climb. Jack is the seventh from the back of the line. How many children are behind Jack? 15 points

$7 - \boxed{1} = \boxed{6}$

⟨Ans.⟩ _____

8 Kate's picture is eighth from the right on the bulletin board. How many pictures are there to the right of Kate's picture? 15 points

⟨Ans.⟩ _____

9 Lucy sings tenth in the school play. How many children sing in front of Lucy? 10 points

⟨Ans.⟩ _____

Standing in line can be tricky!

31

Ordinal Numbers

Date   /   /

Name

Level
★ ★

Score
/ 100

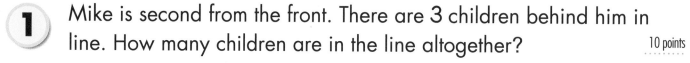

1 Mike is second from the front. There are 3 children behind him in line. How many children are in the line altogether?   10 points

⟨Ans.⟩

2 Nan is fourth from the front of the line to the bathroom. There are 5 children behind her. How many people are waiting for the bathroom in all?   10 points

⟨Ans.⟩

3 We are forming a row for a game. Paul is fourth from the left, and there are 3 children to the right of Paul. How many of us are in the row?   10 points

⟨Ans.⟩

4 Sally is fourth from the back of the line for ice cream. There are 6 people in front of her. How many people are waiting for ice cream in all?   10 points

⟨Ans.⟩

5 Rocky got third in the race, and 5 children placed after him. How many children raced?   10 points

⟨Ans.⟩

6 My comic book is third from the top in a pile of books. There are 4 books under it. How many books are in the pile? 10 points

⟨Ans.⟩ _____

7 Some children are standing in line to turn in their homework. The fifth child finished turning in his work, and there are 4 children left. How many children were there in all? 10 points

⟨Ans.⟩ _____

8 Ted is sixth from the left in the class picture. There are 3 children to the right of him in his row. How many children are in the row? 10 points

⟨Ans.⟩ _____

9 Alik is fourth from the front of the line for tickets, and there are 5 people behind him. How many people are waiting for tickets altogether? 10 points

⟨Ans.⟩ _____

10 Vick is fifth from the front of the line for the bus, and there are 4 people behind him. How many people are waiting for the bus? 10 points

⟨Ans.⟩ _____

Good job, you're doing great!

1 5 children are standing in line for lunch. Andy is second from the front. How many children are behind him? 10 points

⟨Ans.⟩ _____

2 Today, we are getting our flu shots and 8 of us are standing in line. I am fourth from the front. How many people are behind me? 10 points

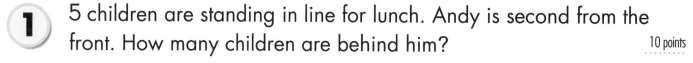

⟨Ans.⟩ _____

3 It's time for the class picture, so all 9 of us are forming a row. Dana is sixth from the left. How many children are to the right of her? 10 points

⟨Ans.⟩ _____

4 8 children are running a race. The third child just finished. How many children are still running? 10 points

⟨Ans.⟩ _____

5 7 people are waiting in line for a train. Dave is fourth from the front. How many people are behind him?

10 points

〈Ans.〉 _____

6 8 people are standing in line for tickets to a movie. Emily is the third from the front. How many people are behind her?

10 points

〈Ans.〉 _____

7 9 people are waiting in line at the store. Nick's father is the second from the back. How many people are in front of him?

10 points

〈Ans.〉 _____

8 There are 8 hats on a row of hooks. Joe's hat is the second from the right. How many hats are there to the left of his hat?

15 points

〈Ans.〉 _____

9 10 people are standing in line for the bus. Harry is sixth from the front. How many people are behind him?

15 points

〈Ans.〉 _____

You're almost there!

Ordinal Numbers

Level ★★   Score   /100

1 5 children are waiting in line. There are 3 children behind Ian. What number child from the front is Ian?   10 points

⟨Ans.⟩

2 8 people are standing in line at a bus stop. There are six people in front of Jack. What number person from the back is Jack?   10 points

⟨Ans.⟩

3 There are 7 books on the bookshelf. There are 4 books to the right of my workbook. What number book from the left is my workbook?   10 points

⟨Ans.⟩

4 6 people are waiting in line for tickets to the show. There are 4 people in front of Kate. What number person from the back is Kate?   10 points

⟨Ans.⟩

5 A train with 6 cars is going to Boston. There are 4 cars in front of the car that Linda is in. What number car from the back of the train is Linda's?   10 points

⟨Ans.⟩

  © Kumon Publishing Co., Ltd.

6 8 books are in a pile. There are 3 books under the sticker book. What number from the top is the sticker book?

10 points

⟨Ans.⟩ _____

7 9 boys are racing in a line. There are 3 boys behind Mack. What number boy from the front is Mack?

10 points

⟨Ans.⟩ _____

8 9 cars are waiting in line for a toll. There are 4 cars behind Nancy's car. What number car from the front is Nancy's?

10 points

⟨Ans.⟩ _____

9 10 children are standing in line for a shot. There are 6 children in front of Paulo. What number child from the back is Paulo?

10 points

⟨Ans.⟩ _____

10 There are 10 pictures in a row on the wall. There are 3 pictures to the right of Betty's picture. What number from the left is Betty's picture?

10 points

⟨Ans.⟩ _____

Are you understanding ordinal numbers? Good!

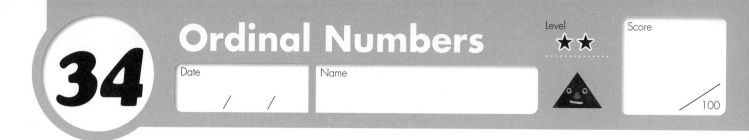

34 **Ordinal Numbers**

Date / /   Name

Level ★ ★

Score /100

1 Some boys are standing in line for their turn. Ricky is fourth from the front. What number from the front is the boy that is second after Ricky?

10 points

〈Ans.〉

2 6 children raced through the park. Susie finished afterward. What number child from the front was she?

10 points

〈Ans.〉

3 Ted is ninth in line for lunch. How many people are in front of him?

10 points

〈Ans.〉

4 Victoria is fifth in line for a ticket, and there are 3 people behind her. How many people are in her line?

10 points

〈Ans.〉

5 10 people are waiting for a bus in line. Wendy is fourth from the back. How many people are in front of Wendy?

10 points

〈Ans.〉

6 There are 8 books on the bookshelf. There are 5 books to the right of the book about bats. What number book from the left is the book about bats?

10 points

〈Ans.〉 _____

7 Andy's picture is sixth from the left in a row on the wall. Bob's picture is third to the right from Andy's. What number picture from the left is Bob's?

10 points

〈Ans.〉 _____

8 Dan and Peggy are on the stairs. Dan is the fourth step from the bottom. Peggy is the sixth step up from Dan. How many steps up from the bottom is Peggy?

10 points

〈Ans.〉 _____

9 10 boys are waiting in line for the bathroom. Ned is the seventh from the front. How many boys are behind Ned?

10 points

〈Ans.〉 _____

10 9 children were waiting in line to buy candy. The fourth child finished. How many children are left in line?

10 points

〈Ans.〉 _____

Wow! You did great.
Now let's review a little.

35 **Review**

Level
★ ★ ★

Date / /

Name

Score
/100

1 There are 4 fish in one tank, and 5 fish in another. How many fish are there in all?

10 points

⟨Ans.⟩

2 You have 8 dimes. If you use 3 dimes, how many are left?

10 points

⟨Ans.⟩

3 Tom had 9 pencils, and then he sharpened 6 of them. How many pencils does Tom have that are not sharp?

10 points

⟨Ans.⟩

4 7 boys and 4 girls are playing tag. Are there more girls or boys? How many more?

10 points

⟨Ans.⟩ There are   more   .

5 I had 9 candies. I gave 2 to my sister and 2 to my brother. How many candies do I have now?

10 points

⟨Ans.⟩

6 There were 8 birds on the roof. Then 6 more landed. How many birds are there altogether?

10 points

⟨Ans.⟩ _____

7 Ellen has 16 marbles. If she gives her sister 7 marbles, how many marbles will remain?

10 points

⟨Ans.⟩ _____

8 There are 12 red fish in the pond. There are 4 fewer goldfish than red fish in the pond. How many goldfish are there?

10 points

⟨Ans.⟩ _____

9 In my garden, I have 12 yellow flowers and 8 pink flowers. How many more yellow flowers than pink flowers do I have?

10 points

⟨Ans.⟩ _____

10 Some children are waiting in line. Fred is ninth from the front, and there are 7 children behind him. How many children are there altogether?

10 points

⟨Ans.⟩ _____

Did you learn everything? Nice job!

36 Review

Level
★ ★ ★

Date
/ /

Name

Score
/100

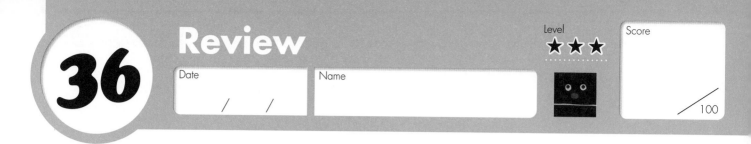

1 6 crows were in the garden. Then 3 more came. How many crows were there in all?

10 points

⟨Ans.⟩ _____

2 I have 6 apples and 8 pears in my basket. How many more pears than apples do I have?

10 points

⟨Ans.⟩ _____

3 Your garden has 7 red flowers and 5 yellow flowers. How many more red flowers than yellow flowers do you have?

10 points

⟨Ans.⟩ _____

4 A bus with 8 people stops at a bus stop, and 2 people get on. 5 people get on at the second stop. How many people are now on the bus?

10 points

⟨Ans.⟩ _____

5 Gina had 10 dimes. After she spent 4 dimes on gum, she got 3 dimes from her sister. How many dimes does she have now?

10 points

⟨Ans.⟩ _____

© Kumon Publishing Co., Ltd.

6 You have 15 stickers. If you use 8 stickers, how many will you have left?

10 points

〈Ans.〉 _____

7 There are 14 fish and 6 turtles in the pond. Are there fewer fish or turtles? How many fewer?

10 points

〈Ans.〉 There are _____ fewer _____ .

8 The teacher has 13 oranges. If he gives 8 children one orange each, how many oranges remain?

10 points

〈Ans.〉 _____

9 13 children are waiting in line. Jim is ninth from the front. How many children are behind Jim?

10 points

〈Ans.〉 _____

10 On Adam's bookshelf, the book about cats is the sixth book from the left. He also has a book about fish that is third on the right of the book about cats. What number book from the left is the book about fish?

10 points

〈Ans.〉 _____

Well done!
Show your parents what you did!

© Kumon Publishing Co., Ltd.

1 Addition
pp 2,3

1. 6 coins
2. 6 balls
3. 5 ducks
4. 6 cats
5. 7 boxes
6. 7 boxes
7. 6 cars
8. 9 birds
9. 9 chicks
10. 6 balloons

2 Addition
pp 4,5

1. 3 + 2 = 5
2. (1) 5 + 3 = 8
   (2) 2 + 5 = 7
   (3) 4 + 2 = 6
   (4) 3 + 4 = 7
3. (1) 4 + 3 = 7
   (2) 3 + 5 = 8
   (3) 2 + 4 = 6
   (4) 4 + 5 = 9
   (5) 6 + 4 = 10

3 Addition
pp 6,7

1. (1) 2 + 3 = 5   **Ans.** 5 birds
   (2) 3 + 4 = 7   **Ans.** 7 apples
   (3) 5 + 3 = 8   **Ans.** 8 fish
   (4) 5 + 4 = 9   **Ans.** 9 eggs

2
(1) 5 + 4 = 9   **Ans.** 9 sheets
(2) 4 + 2 = 6   **Ans.** 6 books
(3) 6 + 3 = 9   **Ans.** 9 pencils
(4) 3 + 2 = 5   **Ans.** 5 eggs
(5) 5 + 3 = 8   **Ans.** 8 children
(6) 5 + 2 = 7   **Ans.** 7 cars

4 Addition
pp 8,9

1. 3 + 6 = 9   **Ans.** 9 flowers
2. 2 + 6 = 8   **Ans.** 8 cookies
3. 4 + 3 = 7   **Ans.** 7 children
4. 6 + 2 = 8   **Ans.** 8 pencils
5. 4 + 2 = 6   **Ans.** 6 balloons
6. 4 + 5 = 9   **Ans.** 9 flies
7. 3 + 2 = 5   **Ans.** 5 balls
8. 5 + 4 = 9   **Ans.** 9 pencils
9. 4 + 3 = 7   **Ans.** 7 oranges
10. 6 + 2 = 8   **Ans.** 8 cars

5 Addition
pp 10,11

1. 2 + 3 = 5
2. 4 + 2 = 6
3. (1) 3 + 4 = 7
   (2) 3 + 2 = 5
4. (1) 3 + 5 = 8
   (2) 5 + 2 = 7
   (3) 4 + 3 = 7
   (4) 6 + 2 = 8
   (5) 7 + 2 = 9
   (6) 4 + 5 = 9

6 Addition
pp 12,13

1
(1) $2 + 4 = 6$ **Ans.**   6 boxes
(2) $3 + 2 = 5$ **Ans.**   5 flowers
(3) $4 + 3 = 7$ **Ans.**   7 candies
(4) $3 + 4 = 7$ **Ans.**   7 turtles

2
(1) $4 + 3 = 7$ **Ans.**   7 cars
(2) $5 + 2 = 7$ **Ans.**   7 birds
(3) $6 + 2 = 8$ **Ans.**   8 children
(4) $7 + 2 = 9$ **Ans.**   9 pencils
(5) $3 + 2 = 5$ **Ans.**   5 dogs
(6) $4 + 5 = 9$ **Ans.**   9 apples

7 Addition
pp 14,15

1 $6 + 2 = 8$ **Ans.**   8 children
2 $4 + 3 = 7$ **Ans.**   7 fish
3 $7 + 2 = 9$ **Ans.**   9 sheets
4 $5 + 3 = 8$ **Ans.**   8 fish
5 $6 + 3 = 9$ **Ans.**   9 comic books
6 $4 + 3 = 7$ **Ans.**   7 hot dogs
7 $5 + 4 = 9$ **Ans.**   9 eggs
8 $4 + 2 = 6$ **Ans.**   6 paper toys
9 $3 + 2 = 5$ **Ans.**   5 oranges
10 $6 + 3 = 9$ **Ans.**   9 sparrows

8 Addition
pp 16,17

1 $5 + 4 = 9$ **Ans.**   9 fish
2 $4 + 6 = 10$ **Ans.**   10 books
3 $6 + 3 = 9$ **Ans.**   9 sheets
4 $6 + 2 = 8$ **Ans.**   8 birds
5 $3 + 4 = 7$ **Ans.**   7 eggs
6 $9 + 3 = 12$ **Ans.**   12 children
7 $7 + 3 = 10$ **Ans.**   10 books
8 $2 + 6 = 8$ **Ans.**   8 ships
9 $8 + 4 = 12$ **Ans.**   12 cards
10 $3 + 4 = 7$ **Ans.**   7 pencils

9 Addition
pp 18,19

1 6 pears
2 7 picture books
3 $4 + 2 = 6$ **Ans.**   6 butterflies
4 $6 + 4 = 10$ **Ans.**   10 chickens
5 $5 + 3 = 8$ **Ans.**   8 girls
6 $3 + 5 = 8$ **Ans.**   8 sheets of blue paper
7 $4 + 2 = 6$ **Ans.**   6 bulbs
8 $5 + 4 = 9$ **Ans.**   9 oranges
9 $7 + 3 = 10$ **Ans.**   10 strawberries

10 Addition
pp 20,21

1 5 cookies
2 5 sheets
3 $3 + 4 = 7$ **Ans.**   7 pencils
4 $4 + 2 = 6$ **Ans.**   6 bells
5 $3 + 5 = 8$ **Ans.**   8 vases
6 $3 + 6 = 9$ **Ans.**   9 cards
7 $4 + 5 = 9$ **Ans.**   9 bags
8 $4 + 4 = 8$ **Ans.**   8 flags
9 $6 + 4 = 10$ **Ans.**   10 chairs

11 Addition
pp 22,23

1 $6 + 2 = 8$ **Ans.**   8 dogs
2 $8 + 4 = 12$ **Ans.**   12 cookies
3 $7 + 4 = 11$ **Ans.**   11 fish
4 $5 + 4 = 9$ **Ans.**   9 children
5 $3 + 5 = 8$ **Ans.**   8 oranges
6 $6 + 2 = 8$ **Ans.**   8 cars

© Kumon Publishing Co., Ltd.

7 $8 + 2 = 10$ **Ans.** 10 picture books

8 $3 + 5 = 8$ **Ans.** 8 paper airplanes

9 $8 + 4 = 12$ **Ans.** 12 children

10 $6 + 3 = 9$ **Ans.** 9 oranges

12 **Subtraction** pp 24,25

1 6 buttons

2 4 boxes

3 1 sparrow

4 3 strawberries

5 4 fish

6 2 apples

7 3 girls

8 4 dogs

9 3 frogs

10 2 textbooks

13 **Subtraction** pp 26,27

1 $3 - 1 = 2$

2 (1) $5 - 2 = 3$

  (2) $6 - 2 = 4$

  (3) $7 - 3 = 4$

  (4) $8 - 2 = 6$

3 (1) $6 - 4 = 2$

  (2) $7 - 2 = 5$

  (3) $8 - 3 = 5$

  (4) $7 - 5 = 2$

  (5) $9 - 4 = 5$

14 **Subtraction** pp 28,29

1 (1) $8 - 2 = 6$ **Ans.** 6 candies

  (2) $6 - 4 = 2$ **Ans.** 2 sheets

  (3) $7 - 3 = 4$ **Ans.** 4 pieces

  (4) $10 - 3 = 7$ **Ans.** 7 fish

2 (1) $6 - 2 = 4$ **Ans.** 4 sparrows

  (2) $10 - 3 = 7$ **Ans.** 7 pencils

  (3) $8 - 5 = 3$ **Ans.** 3 fish

  (4) $7 - 4 = 3$ **Ans.** 3 sheets

  (5) $9 - 3 = 6$ **Ans.** 6 eggs

  (6) $8 - 6 = 2$ **Ans.** 2 flowers

15 **Subtraction** pp 30,31

1 $8 - 5 = 3$ **Ans.** 3 apples

2 $9 - 3 = 6$ **Ans.** 6 cars

3 $7 - 4 = 3$ **Ans.** 3 dollars

4 $8 - 3 = 5$ **Ans.** 5 children

5 $7 - 3 = 4$ **Ans.** 4 balloons

6 $6 - 2 = 4$ **Ans.** 4 picture books

7 $8 - 5 = 3$ **Ans.** 3 pears

8 $7 - 2 = 5$ **Ans.** 5 oranges

9 $9 - 5 = 4$ **Ans.** 4 candles

10 $10 - 2 = 8$ **Ans.** 8 fish

16 **Subtraction** pp 32,33

1 $7 - 3 = 4$ **Ans.** 4 oranges

2 $6 - 2 = 4$ **Ans.** 4 sheets

3 $7 - 3 = 4$ **Ans.** 4 pencils

4 $6 - 2 = 4$ **Ans.** 4 picture books

5 $9 - 4 = 5$ **Ans.** 5 dollars

6 $6 - 2 = 4$ **Ans.** 4 tomatoes

7 $10 - 6 = 4$ **Ans.** 4 balls

8 $7 - 3 = 4$ **Ans.** 4 umbrellas

 © Kumon Publishing Co., Ltd.

9 8 − 3 = 5   **Ans.**   5 yellow balloons

10 8 − 5 = 3   **Ans.**   3 girls

17 **Subtraction** pp 34, 35

1 2 cats

2 7 − 5 = 2   **Ans.**   2 desks

3 8 − 6 = 2   **Ans.**   2 bicycles

4 9 − 5 = 4   **Ans.**   4 red balloons

5 5 − 4 = 1   **Ans.**   1 orange

6 9 − 7 = 2   **Ans.**   2 marbles

7 8 − 3 = 5   **Ans.**   5 frogs

8 7 − 4 = 3   **Ans.**   3 white flowers

18 **Subtraction** pp 36, 37

1 3 apples

2 2 oranges

3 6 − 4 = 2   **Ans.**   2 oranges

4 6 − 2 = 4   **Ans.**   4 ladybugs

5 8 − 5 = 3   **Ans.**   3 monkeys

6 7 − 4 = 3   **Ans.**   3 dishes

7 9 − 6 = 3   **Ans.**   3 boys

8 10 − 4 = 6   **Ans.**   6 marbles

19 **Subtraction** pp 38, 39

1 8 − 6 = 2   **Ans.**   2 apples

2 7 − 4 = 3   **Ans.**   3 apples

3 8 − 5 = 3
   Ans.   I have 3 more apples.

4 9 − 7 = 2
   Ans.   I have 2 more picture books.

5 6 − 4 = 2
   Ans.   There are 2 more dragonflies.

6 5 − 3 = 2
   Ans.   There are 2 fewer blue pencils.

7 8 − 5 = 3
   Ans.   There are 3 fewer pots.

8 10 − 7 = 3
   Ans.   There are 3 fewer hats.

20 **Subtraction** pp 40, 41

1 8 − 3 = 5   **Ans.**   5 oranges

2 6 − 3 = 3   **Ans.**   3 girls

3 7 − 3 = 4   **Ans.**   4 unicycles

4 8 − 2 = 6   **Ans.**   6 pots

5 10 − 2 = 8   **Ans.**   8 yellow balloons

6 9 − 3 = 6   **Ans.**   6 pencils

7 8 − 5 = 3   **Ans.**   3 peaches

8 7 − 2 = 5   **Ans.**   5 red clips

9 9 − 4 = 5   **Ans.**   5 stickers

10 6 − 1 = 5   **Ans.**   5 eggs

21 **Subtraction** pp 42, 43

1 2 chairs

2 2 apples

3 6 − 3 = 3   **Ans.**   3 marbles

4 7 − 5 = 2   **Ans.**   2 cookies

5 8 − 5 = 3   **Ans.**   3 dishes

6 8 − 6 = 2   **Ans.**   2 bells

7 8 − 6 = 2   **Ans.**   2 chairs

8 10 − 8 = 2   **Ans.**   2 oranges

9 9 − 6 = 3   **Ans.**   3 sheets

10 7 − 5 = 2   **Ans.**   2 bottles

(22) Subtraction
pp 44,45

1 $5 - 2 = 3$ **Ans.**   3 apples

2 $8 - 3 = 5$ **Ans.**   5 pencils

3 $6 - 4 = 2$ **Ans.**   2 shells

4 $7 - 3 = 4$ **Ans.**   4 red flowers

5 $9 - 3 = 6$ **Ans.**   6 stickers

6 $8 - 2 = 6$ **Ans.**   6 children

7 $7 - 4 = 3$

   Ans.   There are 3 more boys.

8 $10 - 8 = 2$

   Ans.   Jim jumped 2 more times.

9 $10 - 8 = 2$ **Ans.**   2 times

10 $9 - 6 = 3$ **Ans.**   3 black fish

(23) Addition or Subtraction
pp 46,47

1 $8 - 2 = 6$ **Ans.**   6 sheets

2 $8 + 2 = 10$ **Ans.**   10 sheets

3 $6 + 3 = 9$ **Ans.**   9 years old

4 $6 - 3 = 3$ **Ans.**   3 years old

5 $5 + 4 = 9$ **Ans.**   9 fish

6 $8 - 6 = 2$ **Ans.**   2 peaches

7 $6 + 4 = 10$ **Ans.**   10 shells

8 $9 - 7 = 2$ **Ans.**   2 cows

9 $7 - 2 = 5$ **Ans.**   5 bottles

10 $9 - 6 = 3$

   Ans.   Fred ran 3 more times.

(24) Addition or Subtraction
pp 48,49

1 $7 - 5 = 2$ **Ans.**   2 eggs

2 $7 + 5 = 12$ **Ans.**   12 eggs

3 $9 + 7 = 16$ **Ans.**   16 apples

4 $9 - 7 = 2$ **Ans.**   2 apples

5 $16 - 9 = 7$ **Ans.**   7 girls

6 $9 + 4 = 13$ **Ans.**   13 stickers

7 $9 + 4 = 13$ **Ans.**   13 stickers

8 $6 + 5 = 11$ **Ans.**   11 goldfish

9 $15 - 8 = 7$ **Ans.**   7 cows

10 $14 - 9 = 5$ **Ans.**   5 children

(25) Addition or Subtraction
pp 50,51

1 $7 + 8 = 15$ **Ans.**   15 dimes

2 $13 - 8 = 5$ **Ans.**   5 children

3 $9 + 6 = 15$ **Ans.**   15 cars

4 $9 - 6 = 3$

   Ans.   There are 3 more bikes.

5 $14 - 6 = 8$ **Ans.**   8 red stickers

6 $6 + 8 = 14$ **Ans.**   14 pennies

7 $15 - 9 = 6$ **Ans.**   6 flowers

8 $14 - 6 = 8$ **Ans.**   8 cars

9 $7 + 8 = 15$ **Ans.**   15 chestnuts

10 $15 - 7 = 8$ **Ans.**   8 stamps

(26) Addition or Subtraction
pp 52,53

1 10 people

2 3 people

3 13 people

4 7 people

5 10 tokens

6 7 stickers

(27) Addition or Subtraction
pp 54,55

1 $7 + 3 + 2 = 12$ **Ans.**   12 candies

2 $7 + 3 - 2 = 8$ **Ans.**   8 candies

3 $7 - 3 + 2 = 6$ **Ans.**   6 candies

 © Kumon Publishing Co., Ltd.

4 $7 - 3 - 2 = 2$ **Ans.**   2 candies

5 $9 - 3 - 4 = 2$ **Ans.**   2 nickels

6 $4 - 2 + 8 = 10$ **Ans.**   10 stickers

7 $5 + 3 - 4 = 4$ **Ans.**   4 apples

8 $4 + 6 + 3 = 13$ **Ans.**   13 people

9 $9 - 2 - 3 = 4$ **Ans.**   4 fish

10 $5 - 3 + 8 = 10$ **Ans.**   10 hair clips

(28) Addition or Subtraction   pp 56,57

1 $10 - 4 - 3 = 3$ **Ans.**   3 sheets

2 $8 + 2 - 3 = 7$ **Ans.**   7 children

3 $4 + 6 - 3 = 7$ **Ans.**   7 marbles

4 $5 + 2 + 2 = 9$ **Ans.**   9 birds

5 $8 - 4 + 3 = 7$ **Ans.**   7 people

6 $9 - 3 - 4 = 2$ **Ans.**   2 dogs

7 $4 - 3 + 5 = 6$ **Ans.**   6 seeds

8 $6 - 2 + 5 = 9$ **Ans.**   9 ducks

9 $4 + 3 + 3 = 10$ **Ans.**   10 baskets

10 $5 - 4 + 7 = 8$ **Ans.**   8 stamps

(29) Ordinal Numbers   pp 58,59

1 (1)

(2)

(3)

(4) Ninth

2 $3 + 2 = 5$ **Ans.**   Fifth

3 $5 + 3 = 8$ **Ans.**   Eighth

4 $4 + 5 = 9$ **Ans.**   Ninth

5 $5 + 2 = 7$ **Ans.**   Seventh

6 $4 + 5 = 9$ **Ans.**   Ninth

(30) Ordinal Numbers   pp 60,61

1 (1)

(2) $5 + 1 = 6$ **Ans.**   Sixth

2 $5 + 1 = 6$ **Ans.**   Sixth

3 $6 + 1 = 7$ **Ans.**   Seventh

4 $8 + 1 = 9$ **Ans.**   Ninth

5 $7 + 1 = 8$ **Ans.**   Eighth

6 $6 - 1 = 5$ **Ans.**   5 cars

7 $7 - 1 = 6$ **Ans.**   6 children

8 $8 - 1 = 7$ **Ans.**   7 pictures

9 $10 - 1 = 9$ **Ans.**   9 children

(31) Ordinal Numbers   pp 62,63

1 5 children

2 $4 + 5 = 9$ **Ans.**   9 people

3 $4 + 3 = 7$ **Ans.**   7 children

4 $4 + 6 = 10$ **Ans.**   10 people

5 $3 + 5 = 8$ **Ans.**   8 children

6 $3 + 4 = 7$ **Ans.**   7 books

7 $5 + 4 = 9$ **Ans.**   9 children

8 $6 + 3 = 9$ **Ans.**   9 children

9 $4 + 5 = 9$ **Ans.**   9 people

10 $5 + 4 = 9$ **Ans.**   9 people

(32) Ordinal Numbers   pp 64,65

1 3 children

2 $8 - 4 = 4$ **Ans.**   4 people

3 $9 - 6 = 3$ **Ans.**   3 children

4 $8 - 3 = 5$ **Ans.**   5 children

5 $7 - 4 = 3$ **Ans.**   3 people

6 $8 - 3 = 5$ **Ans.**   5 people

7 $9 - 2 = 7$ **Ans.** 7 people

8 $8 - 2 = 6$ **Ans.** 6 hats

9 $10 - 6 = 4$ **Ans.** 4 people

33 Ordinal Numbers
pp 66,67

1 $5 - 3 = 2$ **Ans.** Second

2 $8 - 6 = 2$ **Ans.** Second

3 $7 - 4 = 3$ **Ans.** Third

4 $6 - 4 = 2$ **Ans.** Second

5 $6 - 4 = 2$ **Ans.** Second

6 $8 - 3 = 5$ **Ans.** Fifth

7 $9 - 3 = 6$ **Ans.** Sixth

8 $9 - 4 = 5$ **Ans.** Fifth

9 $10 - 6 = 4$ **Ans.** Fourth

10 $10 - 3 = 7$ **Ans.** Seventh

34 Ordinal Numbers
pp 68,69

1 $4 + 2 = 6$ **Ans.** Sixth

2 $6 + 1 = 7$ **Ans.** Seventh

3 $9 - 1 = 8$ **Ans.** 8 people

4 $5 + 3 = 8$ **Ans.** 8 people

5 $10 - 4 = 6$ **Ans.** 6 people

6 $8 - 5 = 3$ **Ans.** Third

7 $6 + 3 = 9$ **Ans.** Ninth

8 $4 + 6 = 10$ **Ans.** 10 steps

9 $10 - 7 = 3$ **Ans.** 3 boys

10 $9 - 4 = 5$ **Ans.** 5 children

35 Review
pp 70,71

1 $4 + 5 = 9$ **Ans.** 9 fish

2 $8 - 3 = 5$ **Ans.** 5 dimes

3 $9 - 6 = 3$ **Ans.** 3 pencils

4 $7 - 4 = 3$

Ans. There are 3 more boys.

5 $9 - 2 - 2 = 5$ **Ans.** 5 candies

6 $8 + 6 = 14$ **Ans.** 14 birds

7 $16 - 7 = 9$ **Ans.** 9 marbles

8 $12 - 4 = 8$ **Ans.** 8 goldfish

9 $12 - 8 = 4$ **Ans.** 4 yellow flowers

10 $9 + 7 = 16$ **Ans.** 16 children

36 Review
pp 72,73

1 $6 + 3 = 9$ **Ans.** 9 crows

2 $8 - 6 = 2$ **Ans.** 2 pears

3 $7 - 5 = 2$ **Ans.** 2 red flowers

4 $8 + 2 + 5 = 15$ **Ans.** 15 people

5 $10 - 4 + 3 = 9$ **Ans.** 9 dimes

6 $15 - 8 = 7$ **Ans.** 7 stickers

7 $14 - 6 = 8$

Ans. There are 8 fewer turtles.

8 $13 - 8 = 5$ **Ans.** 5 oranges

9 $13 - 9 = 4$ **Ans.** 4 children

10 $6 + 3 = 9$ **Ans.** Ninth

 © Kumon Publishing Co., Ltd.